Cursive Copy Book

Hymns and Prayers

Cursive Copy Book
Hymns and Prayers

ISBN #: 1-930953-69-0
Memoria Press
www.MemoriaPress.com

Table of Contents

Introduction

What is copybook?

Copybook is a time-honored learning activity in which students copy Scripture, maxims, poetry, and other literature selections. Copybook, like memorization, is a forgotten skill in modern education. However, like memorization, copybook is a beginning skill that is indispensable to the development of good language skills.

What important skills are practiced in copybook? In copybook students practice penmanship, spelling, reading comprehension, punctuation, and vocabulary. Students develop the habits of accuracy, neatness, attention to detail, and patience; they practice correct grammar and a good writing style; they develop an appreciation for what is good, noble, and beautiful in literature.

Copybook is not the only way for the young student to practice each of these skills—but it is the only exercise that develops all of them at the same time. Copybook is concrete and physical; it complements and completes the mental exercise of memorization and reading. Copybook is so simple and obvious that we overlook it.

Even though penmanship is hard work for the young student, he should come to delight in the physical act of creating a beautiful page of writing. Because the student is not burdened by trying to compose his own words, he is liberated to concentrate on copying the immortal words of others. The young student who is not yet able to think great thoughts of his own can grow in wisdom by thinking great thoughts of others.

Teacher Guidelines

The first task of the teacher is to communicate the importance of the lesson to the student and to model enthusiasm. The child should see copybook as an opportunity to give his very best effort in penmanship throughout the year. Students should be inspired to produce quality work because the words they are asked to memorize and copy are special and, in some cases, the sacred words of Scripture. The quality of the work will be a direct reflection of the ability of the teacher to motivate the child to excellence.

The Copybook Lesson outlined below need not be done in one sitting, but may be broken up into two or more sessions. There are 6 skills, or steps, for each lesson:

Step 1

History Time. Write the memory verse on the board. Read the verse in Latin and English. Using "About the Latin Sayings" on page 46 of this book, your *Prima Latina* or *Latina Christiana* Teacher's Manual, or other reference book, discuss the history or significance of the verse or saying. Put the verse or saying in a context that makes it memorable. Some sayings, like *veni, vidi, vici,* lend themselves to great history lessons. Others, like *Salvete, amici Latinae*, are more straightforward and should focus on vocabulary, derivative, and grammar lessons.

Step 2

Language Lesson. Using the blackboard, study the verse or saying with students. Identify unfamiliar words and explain their meanings in context, rather than by using the dictionary. Some of the verses or sayings contain poetic words that are not common in contemporary writing, words such as 'thee,' 'thy,' 'giveth,' etc. Children are intrigued by these 'interesting words,' and early exposure to them enriches their language education. Even though sayings and hymns may also have non-standard punctuation and grammar, the passages can be used to introduce simple grammar topics (nouns, verbs), punctuation, capitalization, plurals, contractions, etc. This is also a great opportunity to parse the Latin words in the verse or saying for part of speech, conjugation, declension, gender, etc.

Step 3

Memorization. Use the disappearing verse technique. Recite the whole verse together several times. Break the verse down into small sections and write them on the board. Have students recite one section together, and then erase that section. Pointing to the 'erased' section on the board, ask students individually and then in unison to recite what you erased. Repeat until the whole verse has been erased from the board. (Knowing the words are going to be erased helps students concentrate and adds excitement.) By the end of the process, students will be able to repeat the whole verse from memory. For instance, for the Table Blessing the sections would be:

Benedic, Domine,
nos et haec
Tua dona,
quae de Tua largitate
sumus sumpturi,
per Christum
Dominum nostrum.
Amen.

Don't forget to follow the Latin exercise with the English translation.

Step 4 ____

Copying. Now that students are familiar with the verse or saying, they copy it in their best handwriting. It is critical that students develop correct techniques in holding the pencil and forming letters. The teacher should monitor students during copybook time and quickly correct any errors in technique. The great value of copybook is that, unlike day-to-day writing, copybook is a special time for students to give their very best effort.

**Some young students may not be able to completely copy verses at the beginning of the year. They should memorize verses according to the schedule but work on alphabet pages for step 4 until they are ready to do copy work.*

Step 5 ____

Proofreading and Correction. Students should proof and correct all work. This is a critical skill that must be taught, encouraged, and practiced from the beginning of the child's education. The proofreading should be kept as a separate exercise with sufficient time devoted to it. The lesson is not complete until this exercise of proofreading and correction produces a 'finished' product.

Step 6 ____

Review. Part of every day should be devoted to review of verses learned. Students love to practice what they already know. Give students the first word or two of each verse and let them complete the verse as a group recitation or individually. As the verses accumulate, students will be amazed at what they know and will develop confidence in their abilities. At the end of the year, students should be able to recite all verses with fluency.

Prayers and Hymns

English versions of the prayers, the song *Gaudeamus Igitur*, and the hymns *Dona Nobis Pacem*, and *Christus Vincit* are given along with the Latin text in cursive for copying. The remaining hymns are given only in Latin; the translations are given with the descriptive material at the end of the section (About the Hymns and Prayers).

Drawing Boxes

There are a few drawing boxes at various places in the book. The student may use these to practice letters or to make a drawing that relates to the material on the page. The teacher may suggest subjects for freehand drawing.

Pages with Blank Lines

Practice pages with blank lines can be found at the end of this book and also on the Memoria Press website at http://www.memoriapress.com/ under 'copybooks'.

Recommended Schedule

Please see complete instructions for each step on previous pages.

Day 1: Step One - History Time (15 min)

Day 2: Step Two - Language Lesson (15 min.)

Day 3: Steps Three & Four - Memory Work and Copying (20 min.)

Day 4: Step Five - Proofreading and Correction (15-30 min)

Day 5: Step Six - Cumulative Review of all Verses (20 min.)

Letter Practice

This Cursive Copybook is not meant to be a detailed study of all aspects of handwriting and penmanship. For more information on these subjects, refer to our
New American Cursive by Iris Hatfield.
The following pages can be used to introduce cursive letter formation and reinforce areas that prove problematic. You may use these exercises to learn and practice cursive letters before beginning copywork. If your student does not need instruction on creating individual letters, you may skip this section and move straight to the first lesson.

Letter Practice
Start on the 'Trace' page to learn the basic strokes for upper and lower case letters. Then move to the 'Trace and Copy' page to practice writing the letters. Always encourage straight lines, round circles, and appropriately sized letters. Once the basic strokes are mastered, your student should be able to make every letter with a little practice. The advantage of cursive over print is that most letters, and even words, can be made in a single continuous stroke.

For a review of basic strokes, use a blank copy sheet in the back to first practice lines:

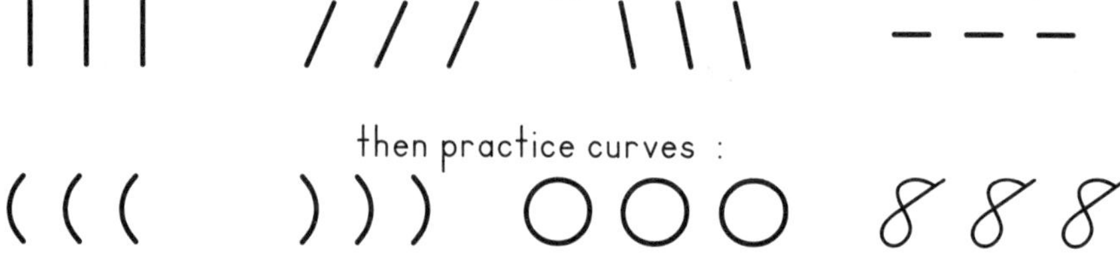

It is not necessary to practice each letter over and over until it is perfect, just until your student has the basic ability to form letters correctly. The purpose of copybook is to allow your student to practice handwriting on great passages, not lists of letters. You will notice significant improvements in letter formation as the year progresses.

Remember, there are several different methods of creating letters. If your child has learned another system that works for him, skip the letter instruction section and move right into the copy work.

Student Guidelines

1. Students should hold pencils properly. The thumb holds the pencil against the second finger with the first finger resting on top.

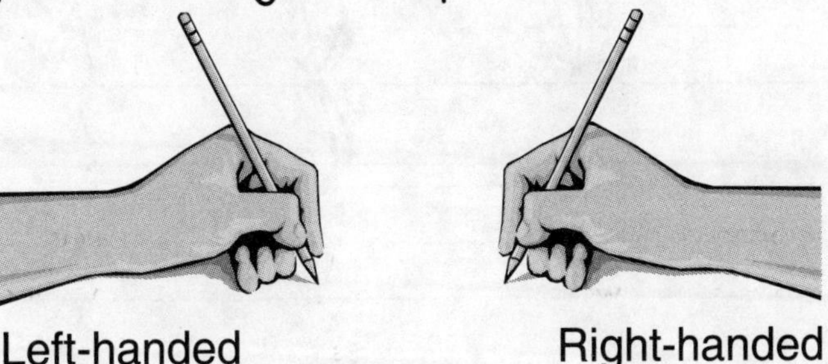

2. Start each line close to the left margin and don't cross the right margin.

3. Uppercase letters must touch the top and third lines.

4. Most lowercase letters stop at the dashed line.

5. Lowercase b, d, f, h, k, l, and t are all tall letters that touch the top line.

6. Lowercase f, g, j, p, q, y, z, and uppercase J and Y have hooks or tails that drop to the bottom line.

7. All words should be one pinky space apart.

8. Letters should be formed at a slight right angle.

9. Erase mistakes completely and be neat.

Trace

Trace and Copy

A

B

C

D

E

F

G

H

I

J

Trace and Copy

K K

L L

M M

N N

O O

P P

Q Q

R R

S S

T T

Trace and Copy

U U

V V

W W

X X

Y Y

Z Z

Trace

a b c d e f

g h i j k l

m n o p q

r s t u v

w x y z

Trace and Copy

a a

b b

c c

d d

e e

f f

g g

h h

i i

j j

Trace and Copy

k k

l l

m m

n n

o o

p p

q q

r r

s s

t t

Trace and Copy

u u

v v

w w

x x

y y

z z

Classroom Latin

Salvete, amici Latinae.

Salvete, amici Latinae.

Salve, magistra.

Salve, magistra.

Salvete, discipuli.

Salvete, discipuli.

Oremus. Repete.

Oremus. Repete.

Quid est tuum praenomen?

Quid est tuum praenomen?

Meum praenomen est ...

Meum praenomen est ...

Classroom Latin

Hello, friends of Latin.

Hello, friends of Latin.

Greetings, teacher.

Greetings, teacher.

Hello, students.

Hello, students.

Let us pray. Repeat.

Let us pray. Repeat.

What is your name?

What is your name?

My name is ...

My name is ...

Classroom Latin

Quid agis? Satis bene.

Quid agis? Satis bene.

Valete, discipuli. Surgite.

Valete, discipuli. Surgite.

Vale, magister. Sedete.

Vale, magister. Sedete.

Deo gratias!

Deo gratias!

Gratias tibi ago.

Gratias tibi ago.

Ego amo te. Optime!

Ego amo te. Optime!

Classroom Latin

How are you? Pretty well.

How are you? Pretty well.

Goodbye, students. Stand up.

Goodbye, students. Stand up.

Goodbye, teacher. Sit down.

Goodbye, teacher. Sit down.

Thanks be to God!

Thanks be to God!

Thank you.

Thank you.

I love you. Excellent!

I love you. Excellent!

Latin Sayings

Ora et labora.

-St. Benedict

Veni, vidi, vici.

-Julius Caesar

Pax Romana

Anno Domini, A.D.

Latin Sayings

Pray and work.

–St. Benedict

I came, I saw, I conquered

–Julius Caesar

The Roman Peace

In the year of our Lord

Latin Sayings

Senatus Populusque

Romanus (S.P.Q.R.)

Romanus civis sum.

- Saint Paul

Caelum et Terra

Stupor mundi

Latin Sayings

The Senate and

the People of Rome

I am a citizen of Rome.

- Saint Paul

Heaven and Earth

Wonder of the world

Latin Sayings

Novus ordo seclorum

Labor omnia vincit.

Nunc et numquam

E pluribus unum

Et tu, Brute?

Mea culpa

Latin Sayings

New order of the ages

Work conquers all.

Now or never

One out of many

You too, Brutus?

My fault, I am guilty

Latin Sayings

Semper fidelis

Excelsior!

Sanctus, Sanctus, Sanctus

Dominus Deus Sabaoth

Mater Italiae - Roma

Alma mater

Latin Sayings

Always faithful

Ever higher!

Holy, Holy, Holy

Lord God of Hosts

The mother of Italy - Rome

Nurturing mother

Latin Sayings

Quo vadis, Domine?

- John 13:36

Rident stolidi verba Latina.

- Ovid

Ante bellum

Signum Crucis

Latin Sayings

Where are you going, Lord?

- John 13:36

Fools laugh at the Latin

language. - Ovid

Before the war

Sign of the Cross

Latin Sayings

Miles Christi sum.

Agnus Dei qui tollis

peccata mundi

Latin Sayings

I am a soldier of Christ.

Lamb of God who takes away

the sins of the world

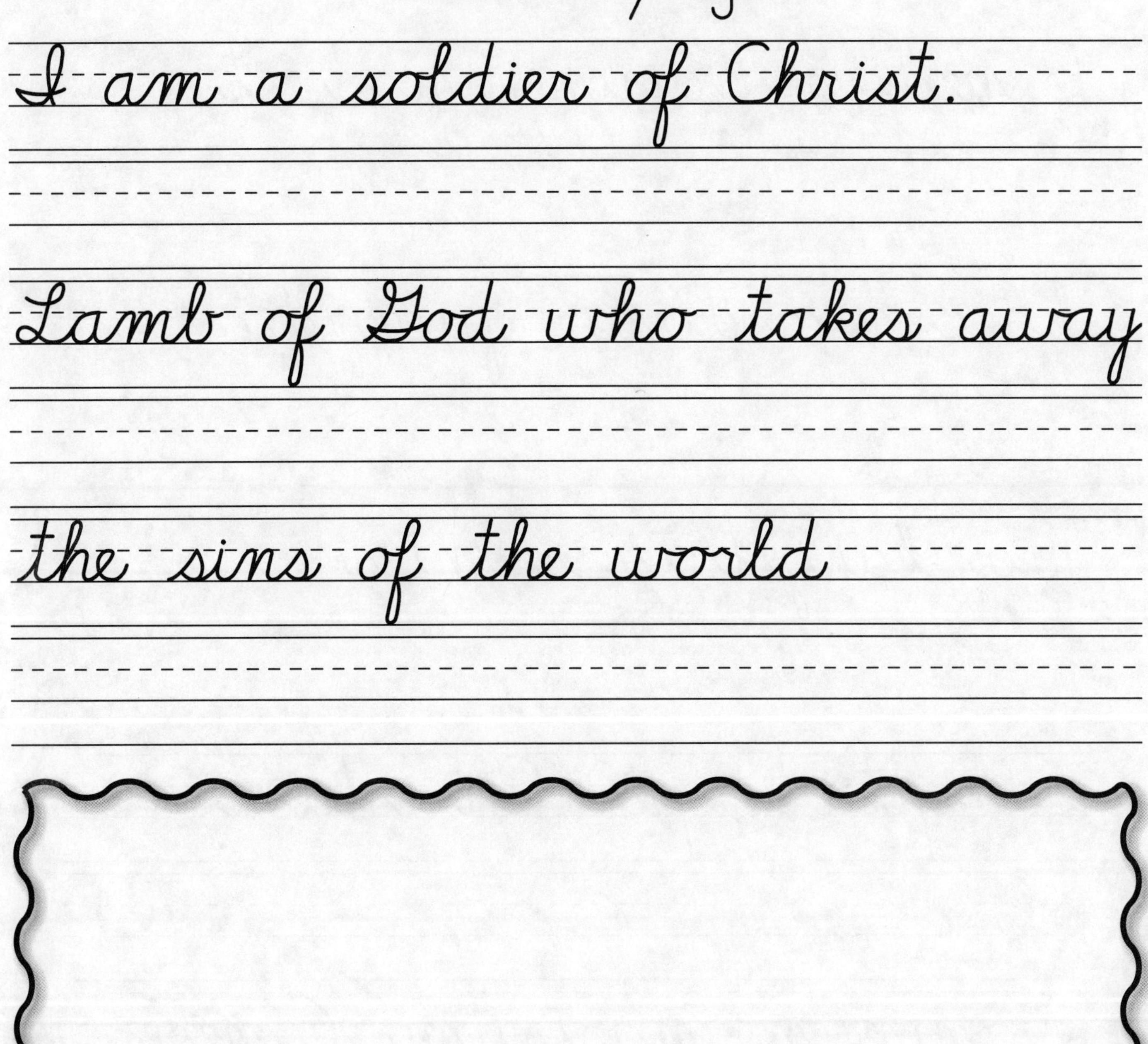

Table Blessing

Benedic, Domine, nos et

haec Tua dona, quae de

Tua largitate sumus

sumpturi, per Christum

Dominum nostrum, Amen

Table Blessing

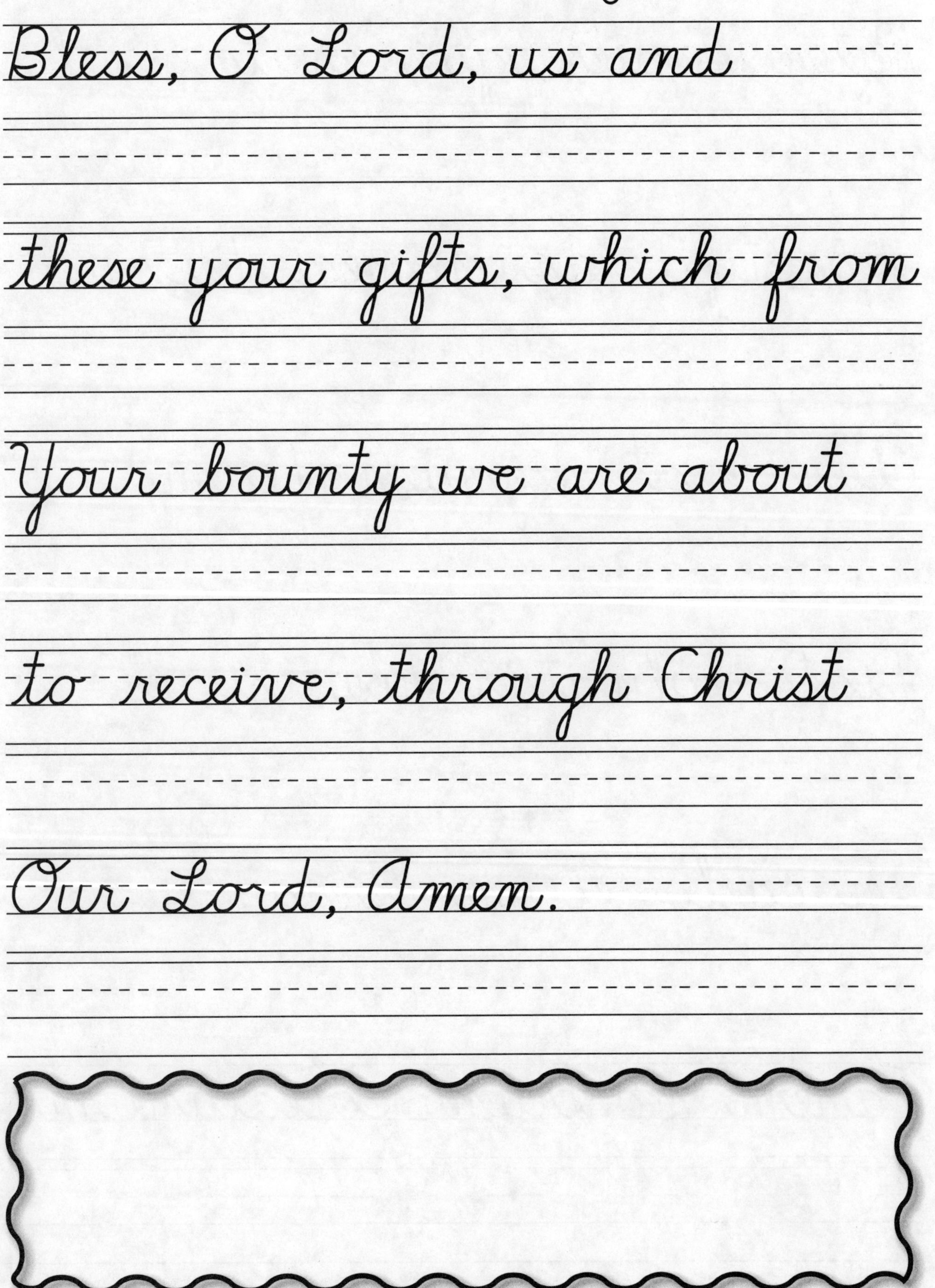

Bless, O Lord, us and

these your gifts, which from

Your bounty we are about

to receive, through Christ

Our Lord, Amen.

The Lord's Prayer

Pater Noster, qui es in caelis,

sanctificetur nomen Tuum.

Adveniat regnum Tuum.

Fiat voluntas Tua, sicut

in caelo et in terra.

Panem nostrum cotidianum

The Lord's Prayer

da nobis hodie, et dimitte

nobis debita nostra sicut

et nos dimittimus

debitoribus nostris. Et ne

nos inducas in tentationem,

sed libera nos a malo.

The Lord's Prayer

Our Father, who art in

Heaven, hallowed be thy

name. Thy kingdom come,

Thy will be done, on earth

as it is in Heaven. Give

us this day our daily bread,

The Lord's Prayer

and forgive us our debts

as we forgive our debtors.

And lead us not into

temptation, but deliver

us from evil.

The Sanctus

Sanctus, Sanctus, Sanctus,

Dominus Deus Sabaoth.

Pleni sunt caeli et terra

gloria Tua. Hosanna in

Excelsis! Benedictus qui

venit in nomine Domini.

The Sanctus

Holy, Holy, Holy, Lord God

of Hosts. Heaven and earth

are full of Your glory.

Hosanna in the highest!

Blessed is he who comes

in the name of the Lord.

The Doxology

Gloria Patri, et Filio,

et Spiritui Sancto,

Sicut erat in principio,

et nunc, et semper,

et in saecula saeculorum

Amen.

The Doxology

Glory be to the Father,

Son, and Holy Spirit.

As it was in the beginning,

it is now and ever

shall be, world without end.

Amen.

Gaudeamus Igitur

Gaudeamus igitur, juvenes

dum sumus. Post jucundam

juventutem, post molestam

senectutem, nos habebit

humus. Vivat academia!

Vivant professores! Vivat

Gaudeamus Igitur

membrum quodlibet! Vivant

membra quaelibet! Semper

sint in flore. Vivant et

res publica, et qui illam

regit. Vivat nostra civitas.

Vivat haec sodalitas quae

Gaudeamus Igitur

nos huc collegit. Alma mater

floreat, quae nos educavit;

caros et commilitones,

dissitas in regiones sparsos

congregavit.

Gaudeamus Igitur

Let us rejoice, therefore, while

we are young. After delightful

youth, after tiresome old age,

the earth will have us.

Long live the academy!

Long live the professors!

Gaudeamus Igitur

Long live the grads! Long
live the undergrads! May
they ever flourish. Long
live the republic and he
who rules it. Long live our
town folk. Long live this

Gaudeamus Igitur

association that gathers us

here. May our Alma Mater

thrive; she has educated us.

She has brought together

friends and colleagues, now

widely scattered.

Dona Nobis Pacem

Dona nobis pacem.

Christus Vincit

Christus vincit,

Christus regnat,

Christus imperat.

Dona Nobis Pacem

Give us peace.

Christus Vincit

Christ conquers,

Christ reigns,

Christ rules.

Veni Creator Spiritus

Veni, Creator Spiritus,

mentes tuorum visita.

Imple superna gratia

quae tu creasti pectora.

Qui diceris Paraclitus,

altissimi donum Dei,

Veni Creator Spiritus

fons vivus, ignis, caritas,

et spiritalis unctio.

Tu, septiformis munere,

digitus paternae dexterae,

Tu rite promissum Patris,

sermone ditans guttura.

Veni Creator Spiritus

Accende lumen sensibus;

infunde amorem cordibus;

infirma nostri corporis

virtute firmans perpeti.

Hostem repellas longius,

pacemque dones protinus;

Veni Creator Spiritus

ductore sic te praevio

vitemus omne noxium.

Per te sciamus da Patrem,

noscamus atque Filium;

Teque utriusque Spiritum

credamus omni tempore.

Adeste Fideles

Adeste, fideles,

Laeti triumphantes.

Venite, venite in Bethlehem.

Natum videte

Regem angelorum.

Venite adoremus, Dominum.

Adeste Fideles

Deum de Deo,

lumen de lumine,

gestant puellae viscera.

Deum verum,

genitum, non factum.

Venite adoremus, Dominum.

Veni, Veni Emmanuel

Veni, veni Emmanuel,

captivum solve Israel,

qui gemit in exilio,

privatus Dei Filio.

Gaude! Gaude! Emmanuel

nascetur pro te Israel.

Veni, Veni Emmanuel

Veni, veni Rex Gentium

Veni Redemptor omnium,

ut salvas tuos famulos

peccati sibi conscios.

Veni, veni O Oriens,

solare nos adveniens,

Veni, Veni Emmanuel

noctis depelle nebulas

dirasque mortis tenebras.

Veni, Clavis Davidica,

regna reclude caelica,

fac iter tutum supernum,

et claude vias inferum.

Veni, Veni Emmanuel

Veni, O Iesse Virgula,

ex hostis tuos ungula,

de spectu tuos tartari

educ et antro barathri.

Veni, veni Adonai,

qui populo in Sinai

Veni, Veni Emmanuel

legem dedisti vertice

in majestate gloriae.

Veni, O Sapientia,

quae hic disponis omnia,

veni, viam prudentiae

ut doceas et gloriae.

Resonet in Laudibus

Resonet in laudibus

cum jucundis plausibus

Sion cum fidelibus.

Apparuit quem genuit Maria.

Gaudete, gaudete,

Christus natus hodie

Resonet in Laudibus

ex Maria Virgine.

Sion lauda Dominum

Salvatorem omnium.

Virgo parit Filium.

Pueri concurrite,

nato Regi psallite,

voce pia dicite.

Natus est Emmanuel

Quem praedixit Gabriel.

Testis est Ezechiel.

Ubi Caritas

Ubi caritas et amor,

Deus ibi est.

Congregavit nos in unum

Christi amor. Exultemus,

et in ipso jucundemur.

Timeamus, et amemus

Ubi Caritas

Deum vivum, et ex corde

diligamus nos sincero.

Simul ergo cum in unum

congregamur, ne nos mente

dividamur, caveamus.

Cessent iurgia maligna,

Ubi Caritas

cessent lites. Et in medio

nostri sit Christus Deus.

Simul quoque cum beatis

videamus, glorianter

vultum tuum, Christe Deus.

Gaudium quod est

Ubi Caritas

immensum atque probum,

saecula per infinita

saeculorum. Amen.

Pange Lingua

Pange lingua gloriosi

corporis mysterium,

sanguinisque pretiosi,

quem in mundi pretium

fructus ventris generosi

Rex effudit gentium.

Pange Lingua

Nobis datus, nobis natus

ex intacta Virgine,

et in mundo conversatus,

sparso verbi semine,

sui moras incolatus

miro clausit ordine.

Pange Lingua

In supremae nocte caenae

recumbens cum fratribus,

observata lege plene,

cibis in legalibus,

cibum turbae duodenae,

se dat suis manibus.

Pange Lingua

Verbum caro, panem verum,

verbo carnem efficit;

fitque sanguis Christi merum,

et si sensus deficit,

ad firmandum cor sincerum,

sola fides sufficit.

Tantum Ergo

Tantum ergo Sacramentum

veneremur cernui;

et antiquum documentum

novo cedat ritui;

praestet fides supplementum

sensum defectui.

Tantum Ergo

Genitori, Genitoque

laus et jubilatio,

salus, honor, virtus quoque

sit et benedictio;

procedenti ab utroque

compar sit laudatio.

Panis Angelicus

Panis angelicus fit panis

hominum, dat panis

coelicus figuris terminum.

O res mirabilis!

manducat Dominum

pauper, servus, et humilis.

Panis Angelicus

Te trina Deitas unaque

poscimus, sic nos tu visita

sicut te colimus, per tuas

semitas duc nos quo

tendimus, ad lucem

quam inhabitas. Amen.

Ave Verum Corpus

Ave verum Corpus natum

de Maria Virgine;

vere passum, immolatum

in cruce pro homine.

O Jesu dulcis! O Jesu pie!

O Jesu, fili Mariae!

Ave Verum Corpus

Cujus latus perforatum

fluxit aqua et sanguine.

Esto nobis praegustatum

mortis in examine.

O Jesu dulcis! O Jesu pie!

O Jesu, fili Mariae!

Adoro Te Devote

Adoro te devote, latens

Deitas, quae sub his figuris

vere latitas; tibi se cor

meum totum subjicit, quia

te contemplans totum deficit.

Visus, tactus, gustus in te

fallitur, sed auditu solo

tuto creditur; credo quidquid

dixit Dei Filius: nil hoc

verbo Veritatis verius. In

cruce latebat sola Deitas,

at hic latet simul et

Adoro Te Devote

humanitas; ambo tamen

credens atque confitens, peto

quod petivit latro paenitens.

Plagas, sicut Thomas non

intueor; Deum tamen meum

te confiteor; fac me tibi

Adoro Te Devote

semper magis credere, in te

spem habere, te diligere.

O memoriale mortis Domini!

Panis vivus, vitam

praestans homini! Praesta

meae menti de te vivere et

Adoro Te Devote

te illi semper dulce sapere.

Pie pellicane, Jesu Domine,

me immundum munda tuo

sanguine; cujus una stilla

salvum facere totum

mundum quit ab omni

Adoro Te Devote

scelere. Iesu, quem velatum

nunc aspicio, oro fiat illud

quod tam sitio; ut te

revelata cernens facie visu

sim beatus tuae gloriae.

Salve, Mater Misericordiae

Salve, Mater misericordiae,

Mater Dei et Mater veniae,

Mater spei et Mater gratiae,

Mater plena

sanctae laetitiae,

O Maria!

Salve, Mater Misericordiae

Salve, felix Virgo puerpera;

nam qui sedet in Patris

dextera, caelum regens,

terram et aethera, intra

tua se clausit viscera,

O Maria!

Salve, Mater Misericordiae

Esto Mater, nostrum solatium,

nostrum esto tu virgo

gaudium, et nos tandem

post hoc exilium, laetos

iunge choris caelestium,

O Maria!

Salve Regina

Salve Regina,

Mater misericordiae.

Vita, dulcedo,

et spes nostra, salve.

Ad te clamamus,

exsules, filii Hevae.

Salve Regina

Ad te suspiramus,

gementes et flentes

in hac lacrimarum valle.

Eia ergo, Advocata nostra,

illos tuos misericordes

oculos ad nos converte.

Salve Regina

Et Jesum, benedictum

fructum ventris tui,

nobis post hoc

exsilium ostende.

O clemens, O pia,

O dulcis Virgo Maria.

Gloria

Gloria in excelsis Deo,

et in terra pax hominibus

bonae voluntatis. Laudamus

te, benedicimus te, adoramus

te, glorificamus te. Gratias

agimus tibi propter magnam

Gloria

gloriam tuam. Domine

Deus, Rex caelestis, Deus

Pater omnipotens. Domine

Fili unigenite, Jesu Christe.

Domine Deus, Agnus Dei,

Filius Patris. Qui tollis

Gloria

peccata mundi, miserere

nobis. Qui tollis peccata

mundi, suscipe deprecationem

nostram. Qui sedes ad

dexteram Patris, miserere

nobis. Quoniam tu solus

Gloria

Sanctus. Tu solus Dominus.

Tu solus Altissimus Jesu

Christe. Cum Sancto

Spiritu, in gloria Dei Patris.

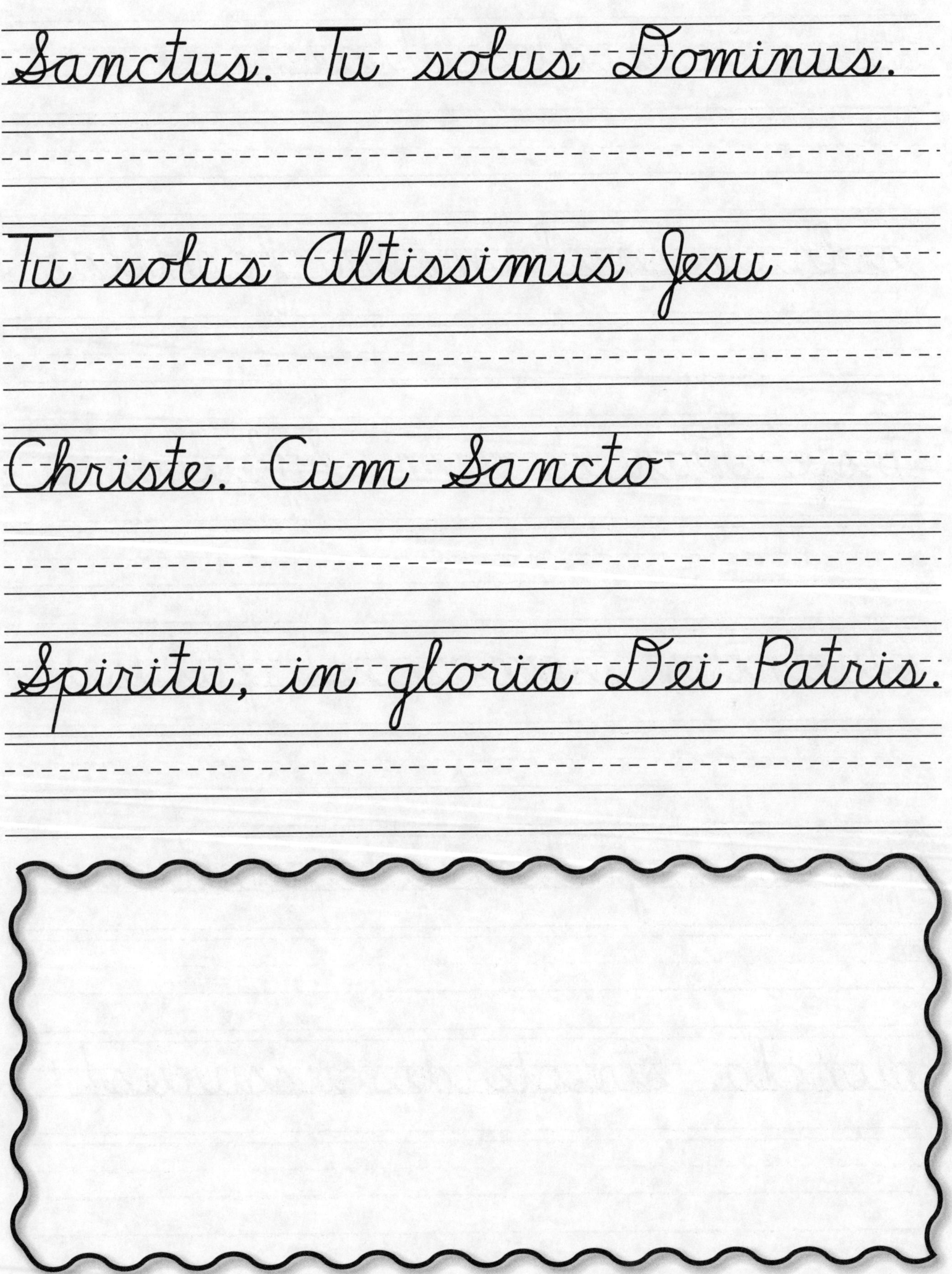

Dies Irae

Dies irae dies illa

solvet saeclum in favilla:

teste David cum Sybilla.

Quantus tremor est futurus

quando judex est venturus

cuncta stricte discussurus!

Dies Irae

Tuba mirum spargens sonum

per sepulchra regionum

coget omnes ante thronum.

Mors stupebit et natura,

cum resurget creatura,

judicanti responsura.

In Paradisum

In paradisum deducant te

angeli; in tuo adventu

suscipiant te martyres, et

perducant te in civitatem

sanctam Jerusalem. Chorus

angelorum te suscipiat, et

In Paradisum

cum Lazaro quondam

paupere aeternam habeas

requiem.

About the Latin Sayings

p. 22: Ora et Labora (Pray and Work)
Ora et labora is a summary of St. Benedict's famous Rule for the monks at his monastery.

p. 22: Veni, vidi, vici. (I came, I saw, I conquered.)
After winning a war in Egypt, Caesar went to Asia Minor and won a victory over the King of Pontus. *Veni, vidi, vici* is the brief dispatch he sent to the Roman Senate in 47 B.C. reporting this victory.

p. 22: Pax Romana (The Roman Peace)
The 'Roman Peace' was the long period of peace and prosperity enjoyed by the inhabitants of the Roman Empire during the 1st and 2nd centuries, beginning with the reign of Augustus.

p. 22: Anno Domini (A.D.) (In the Year of Our Lord)
This designation was invented in the Middle Ages as a way of numbering years beginning with the birth of Christ. It is the same formula the Romans used to number years beginning with the founding of the city (*AUC—anno urbis conditae*, 'in the year of the founded city')

p. 24: Senatus Populusque Romanus (S.P.Q.R.) (The Senate and the People of Rome)
S.P.Q.R. is the symbol of the Roman Republic, written on monuments and official documents.

p. 24: Romanus civis sum. (I am a citizen of Rome.)
This is a formula by which a Roman citizen asserted his rights and privileges. St. Paul used this statement to claim his right as a Roman citizen to be tried at Rome.

p. 24: Caelum et Terra (Heaven and Earth)
This just means 'everything and everywhere.'

p. 24: Stupor mundi (Wonder of the world)
This a list of the seven wonders of the ancient world composed by the geographers of Alexandria.

p. 26: Novus ordo seclorum (New order of the ages)
This motto, declaring a new age based on reason, appears on the U.S. one-dollar bill.

p. 26: Labor omnia vincit (Work conquers all)
A paraphrase of a line from Virgil's *Aeneid.*

p. 26: Nunc aut numquam (Now or never)
An ancient saying whose origin is unknown.

p. 26: E pluribus unum (One out of many)
The motto of the United States, referring to one nation made up of many states.

p. 26: Et tu, Brute? (You too, Brutus?)
Words spoken by the character Julius Caesar in Shakespeare's play of the same name.

About the Latin Sayings

p. 26: Mea culpa (My fault)
A common expression, taken from the Confiteor of the Latin Mass, in which both the priest and the people acknowledge sin before coming to the altar.

p. 28: Semper fidelis (Always faithful)
The motto of the United States Marine Corps.

p. 28: Excelsior! (Ever higher!)
The motto of the state of New York.

p. 28: Sanctus, Sanctus, Sanctus, Dominus Deus Sabaoth (Holy, Holy, Holy, Lord God of Hosts)
The Sanctus is one of the oldest recorded elements of the worship liturgy of the Christian church.

p. 28: Mater Italiae—Roma (Mother of Italy—Rome)
From the Roman historian Florus.

p. 28: Alma mater (Nurturing mother)
John Cardinal Newman's description of what a university should be.

p. 30: Quo vadis, Domine? (Where are you going, Lord?)
The question Peter asks when Jesus tells his disciples that he will not be with them much longer.

p. 30: Rident stolidi verba Latina. (Fools laugh at the Latin language.)
A quotation from the Augustan-age Roman poet Ovid.

p. 30: Ante bellum (Before the war)
A phrase usually used in reference to the period preceding the American Civil War.

p. 30: Signum Crucis (Sign of the Cross)
Making the sign of the cross is practiced today by members of the Roman Catholic and Eastern Orthodox Churches.

p. 32: Miles Christi sum. (I am a soldier of Christ.)
In Eph. 6:11-17, St. Paul uses the armor of the Roman soldier as a metaphor for the spiritual armor of the Christian soldier.

p. 32: Agnus Dei qui tollis peccata mundi (Lamb of God who takes away the sins of the world)
An ancient part of church worship, these are the words of John the Baptist proclaiming the Messiah to the Jews in John 1:29.

Additional information about these Sayings can be found in Latina Christiana I *(Memoria Press).*

About the Prayers and Hymns

p. 34: Table Blessing (Grace)
St. Paul counseled the faithful: "Whether you eat or drink, or whatsoever else you do, do all to the glory of God" (1 Corinthians 10:31). It is a custom in most religions to say a short prayer of thanks before a meal. *Benedic, Domine* is a widely-used Catholic grace.

p. 36: The Lord's Prayer (Pater Noster)
The Lord's Prayer is probably the best-known prayer in Christianity. The Pater Noster is the traditional Catholic version, but there are many others. Two versions of it occur in the New Testament, one in Matthew 6:9-13 and the other in Luke 11:2-4.

p. 40: The Sanctus
The Sanctus is the last part of the Preface in the Mass, sung in practically every rite by the people (or choir). It is one of the elements of the liturgy of which we have the earliest evidence.

p. 42: The Doxology (Gloria Patri)
A doxology (from the Greek *doxa*, 'glory' + *logos*, 'word' or 'speaking') is a short hymn of praise to God in various Christian worship services. Among Christian traditions, a doxology is typically a sung expression of praise to the Holy Trinity—the Father, the Son, and the Holy Spirit. The Gloria Patri is commonly used as a doxology by Catholics, Orthodox, Anglicans, and many Protestants.

p. 44: Gaudeamus Igitur
De Brevitate Vitae (*On the Shortness of Life*), more commonly known by its first words *Gaudeamus Igitur*, is a popular academic drinking song in many European countries. In modern times, it is sung at graduation ceremonies. The melody was adapted around 1270 from a medieval hymn by Strada, bishop of Bologna. The song is often sung in a prank version with undignified lyrics.

p. 50: Dona Nobis Pacem and Christus Vincit
Dona Nobis Pacem is a phrase in the *Agnus Dei* section of the Roman Catholic mass. It was set as a separate, final movement in Bach's Mass in B Minor. It has become a popular piece for children's choirs and is delightful when sung as a round.
Christus Vincit is an 8th-century Ambrosian chant. It has a simple regal melody with simple words for beginners.

p. 52: Veni Creator Spiritus
This serene and solemn hymn is used at Vespers, Whitsunday (Pentecost), dedications, confirmations, and whenever the Holy Spirit is invoked. It is attributed to Rabanus Maurus, Archbishop of Mainz (d. 856). It was used in the contemporary movie *Joan of Arc*. The combination of the music and the words powerfully convey the presence of the Holy Spirit. *Veni Creator Spiritus* is a traditional hymn for the opening of the school year at academic institutions and would be a wonderful tradition to resurrect.

Translation: 1) Come, Creator Spirit, visit the souls of thy people; fill with heavenly grace the hearts which thou hast created. 2) Thou who art called 'Comforter,' the gift of God most high, the living fountain, fire, love, and spiritual balm. 3) Thou, with sevenfold gifts, the finger of the Father's right hand, Thou the promise by rite of the Father, enriching (our) throats with speech. 4) Light a torch to (our) senses, pour love into (our) hearts; keep strengthening with virtue the weaknesses of our bodies. 5) Mayest Thou drive the enemy far (from us), give (us) peace right away. Thus with Thee as (our) guide before (us), may we avoid all things harmful.

p. 56: Adeste Fideles (O Come, All Ye Faithful)
The tune was written by John Francis Wade in 1743. The authorship of the original four verses of the text is uncertain—it may be older or may have been written by Wade himself. *Adeste Fideles* is one of the most popular carols; it has been adapted many times and has been translated into dozens of languages. This carol is the next-to-last hymn sung at the Festival of Nine Lessons and Carols in King's College, Cambridge.

Translation: 1) O come, all ye faithful, joyful and triumphant. O come ye, o come ye to Bethlehem. Come and adore Him, born the King of angels. O come let us adore him, Christ the Lord. 2) The flesh of a virgin bears God from God, light from light. True God begotten, not made.

p. 58: Veni Veni Emmanuel
This hymn actually traces its origins to the church liturgy prior to the ninth century. The text is derived from the seven great *O Antiphons*, which were said at Vespers during the octave before Christmas (Dec. 17-23): *O Emmanuel, O King of All Nations, O Dayspring, O Key of David, O Rod of Jesse, O Adonai, O Wisdom.* The hymn itself was composed in the 12th century in French and the Latin version of the hymn is from the 18th century. Today it is a popular Christmas carol in both English and Latin.

Translation: 1) Come, come Emmanuel, free captive Israel that laments in exile deprived of the Son of God. Rejoice, rejoice! Emmanuel is born for thee, Israel. 2) Come, come, king of nations. Come, Redeemer of all, that you may save your subjects (who are) aware of their own error. 3) Come, come O Rising Sun. Comfort us in coming. Dispel the clouds of night and the fearful shadows of death. 4) Come, key of David. Open up the heavenly kingdoms. Make the journey above safe and close the ways below. 5) Come, O Branch of Jesse. Lead out Thy children from the enemy's claw. Lead out Thy children from the vision of hell. Lead them out also from the pit of the underworld. 6) Come, come Adonai, who to the people on Sinai gave the law at the summit in the majesty of glory. 7) Come, O Wisdom, who ordains all things here (on earth). Come, so that You may show the way of knowledge and of glory.

p. 64: Resonet in Laudibus
Resonet in Laudibus is a 14th-century Latin carol from Germany. Its tune was later adapted for a lullaby by the Virgin Mary in a mystery play. The translation of this lullaby, *Joseph Dearest*, is the text found in most songbooks today. *Resonet in Laudibus* is very easy to sing, a good piece for beginners, and a good lullaby to sing to little ones to start them out in Latin early.

Translation: 1) Let Zion with the faithful resound in praises and joyful clapping. He whom Mary has borne is made known. Rejoice, rejoice, Christ is born today from the Virgin Mary. 2) Zion, praise the Lord, the Savior of all. The Virgin brings forth the Son. 3) Come together, children, sing to the newborn king, speak with a sweet voice. 4) Emmanuel is born, whom Gabriel foretold. Ezekiel is a witness.

About the Prayers and Hymns

p. 66: Ubi Caritas
A popular hymn for Holy Thursday.

Translation: Intro) Where charity and love are, God is there. 1) Christ's love has gathered us into one. Let us rejoice, and let us be pleased in Him. Let us fear, and let us love the living God, and let us love out of a sincere heart. 2) Now when we are gathered at the same time in one, let us beware that we not be divided in mind. Let the spiteful conflicts cease, let the quarrels cease, and let Christ God be in the midst of us. 3) Also, at the same time, with the blessed let us see Thy face in greater glory, Christ God, the joy that is beyond measure and honest, through infinite ages of ages.

p. 70: Pange Lingua
"Sing, my tongue, of the mystery of the glorious body and precious blood." The most commonly sung Corpus Christi hymn and one of the Seven Great Hymns of the Church. The rhythm of the *Pange Lingua* is said to have come down from a marching song of Caesar's legions: "Ecce, Caesar nunc triumphat qui subgegit Gallias."

Translation: 1) Compose, tongue, the mystery of the glorious body and of the precious blood which, in reward of mankind, the fruit of the noble womb, the King of all nations, poured out. 2) Give to us, born to us out of the untouched Virgin, and having lived in the world, having cultivated the battalions with the sown seed of His own word, he ended with wonderful propriety.

p. 74: Tantum Ergo
For many years the Church required that the last two verses of the *Pange Lingua*, beginning *Tantum Ergo*, be sung any time the Blessed Sacrament was exposed. *Pange Lingua* and *Tantum Ergo* are also sung on Holy Thursday.

Translation: 1) Therefore we, bowed down, venerate the Sacrament so much; and may the ancient example give way to the new rite; may faith provide a supplement, understanding for (our) weakness. 2) To the Father and to the Son, praise and jubilation; salvation, honor, virtue also, and may there be blessing; to that proceeding from both, may there be equal praise.

p. 76: Panis Angelicus
Panis Angelicus is the penultimate strophe of the hymn *Sacris Solemniis*, written by St. Thomas Aquinas for the Feast of Corpus Christi.

Translation: 1) The angelic bread becomes the bread of men. Heavenly bread gives boundary to forms—O amazing thing! The pauper, slave, and humble man devours the Lord. 2) Together we ask Thee, threefold Deity; so visit us, so likewise let us worship Thee; by the paths, lead us to where we strive, to the light which You inhabit.

About the Prayers and Hymns

p. 78: Ave Verum Corpus

A short Eucharistic hymn attributed to Pope Innocent VI in the 14th century. It is very beautiful, simple to sing and to translate.

Translation: 1) Hail true body born of the Virgin Mary; truly suffered and sacrificed on the cross for man. 2) Whose pierced side flowed with water and blood, be for us a foretaste of death in judgment.

p. 80: Adoro Te Devote

There is little hard evidence that St. Thomas wrote this hymn. The vocabulary is his but the rhythm is looser, more like a meditation. There is a tradition that it was his last prayer after his last communion, a sentiment so appealing that no scholar poking around for 'facts' will probably ever unseat it.

Translation: 1) I worship Thee devotedly, hidden Deity, who beneath these forms truly hide; to thee my whole soul subjects itself, because contemplating thee, all of it falls short. 2) Sight, touch, taste, with respect to Thee may be faulty, but in fact by hearing only is it safely believed. I believe each thing the Son told of God. Nothing is truer than this word of Truth. 3) On the cross was concealed the one Deity, and here is concealed at the same time humanity also. Yet both believing and acknowledging, I ask that which the penitent thief asked. 4) Just as Thomas, I do not look at the wounds. Still, my God, I admit Thee. Make me always believe in Thee more, keep hope in Thee, love Thee. 5) O memorable thing, the death of the Lord, bread of life bestowing life to man, bestow to my mind to live in accordance with Thee and always to taste Thee sweetly. 6) Holy Pelican, Lord Jesus, cleanse my impure self with Thy blood, of which a single drop can make the whole world saved from every sin. 7) Jesus, whom veiled I now behold, I pray that that which I thirst for so much may happen; that discerning Thee with Thy face revealed, I might be blessed by the vision of Thy glory.

p. 86: Salve, Mater Misericordiae

An ancient Carmelite hymn, with an exceedingly simple melody for students to sing.

Translation: 1) Greetings, Mother of mercy, Mother of God and Mother of kindness, Mother of hope and Mother of grace, Mother full of holy joy, Oh Mary. 2) Greetings, fruitful child-bearing Virgin: for you who sit on the right side of the Father, ruling Heaven, earth, and the aether, He has enclosed Himself within thy womb. 3) You shall be, Mother, our solace; you, Virgin, shall be our joy. And at last, after this exile, join us, joyful, to the choruses of the angels.

p. 89: Salve Regina

One of the most celebrated of all the Marian hymns, **Salve Regina** has long been associated with St. Bernard of Clairvaux, that great saint of the Middle Ages, famous for his reform of the monastic life at Cluny, and for his devotion to the Virgin Mary. Today it is ascribed to another writer, but it is still chanted every day in the cathedral of Speyer in honor of St. Bernard.

Translation: 1) Greetings, Queen, Mother of mercy; our life, sweetness, and hope, greetings. To thee we wanderers, sons of Eve, cry out. 2) To thee we sigh, moaning and weeping, in this valley of tears. Ah, then, our Counselor, turn those (thy) merciful eyes toward us. 3) Thy merciful eyes turn toward us. And Jesus, blessed fruit of Thy womb… 4) reveal to us, after this exile. O merciful, O holy, O sweet Virgin Mary.

About the Prayers and Hymns

p. 92: Gloria

After asking God for mercy, praise is in order. The Gloria is the Church's great hymn of praise to God and the Trinity. It is very ancient, very simple, and very direct. The opening comes from Luke 2:14, where the angels announce the birth of Christ to the shepherds.

Translation: Glory to God on high, and on earth peace to men of good will. We praise you, we bless you, we adore you, we glorify you. We give you thanks for your great glory. Lord God, Heavenly King, God the Father Almighty. Lord Jesus Christ, only begotten Son, Lord God, Lamb of God, Son of the Father. You who take away the sins of the world, have mercy on us. You who take away the sins of the world, receive our prayer. You who sit at the right hand of the Father, have mercy on us. For you alone are holy, you alone are the Lord. You alone are the most high, Jesus Christ. With the Holy Spirit in the glory of God the Father. Amen.

p. 96: Dies Irae

Dies Irae is considered by many to be the greatest Latin hymn ever written. It has been translated hundreds of times into English alone by such greats as Dryden, Scott, and Macaulay. It is a sequence of 18-20 stanzas for the Mass of the Dead. The most famous musical setting is Mozart's *Requiem* Mass. The work is taken from Zephaniah 1:15 : "That day is a day of wrath, a day of trouble and distress, a day of wasteness and desolation, a day of darkness and gloominess, a day of clouds and thick darkness." *Dies Irae* is a profound meditation on the Last Judgment, when Christ will come for the second time, not in mercy, but in judgment.

Translation: 1) Day of wrath, that day will bring the world to an end in ash: David together with Sybil as witness. 2) How much trembling there will be when the judge shall come to examine one and all rigorously. 3) The trumpet spreading a marvelous sound throughout the tombs of the regions shall gather all before the throne. 4) Death will be astonished and nature too when a creature will rise again to answer to the judging. 5) The written book shall be brought forth, in which everything is contained, from which mankind should be judged. 6) Then, when the judge shall take his seat, whatever is hidden will be revealed; nothing will remain unpunished.

p. 98: In Paradisum

The farewell prayer of the *Requiem* mass. A beautiful and fitting end, it should be the consolation of every Christian funeral.

Translation: Into paradise may the angels lead thee; on thy arrival may the martyrs receive thee and guide thee into the holy city of Jerusalem. May a chorus of angels receive thee, and with Lazarus the pauper, hereafter may you have eternal rest.